A LOOK AT THE HUMAN BODY

THE DIGESTIVE SYSTEM

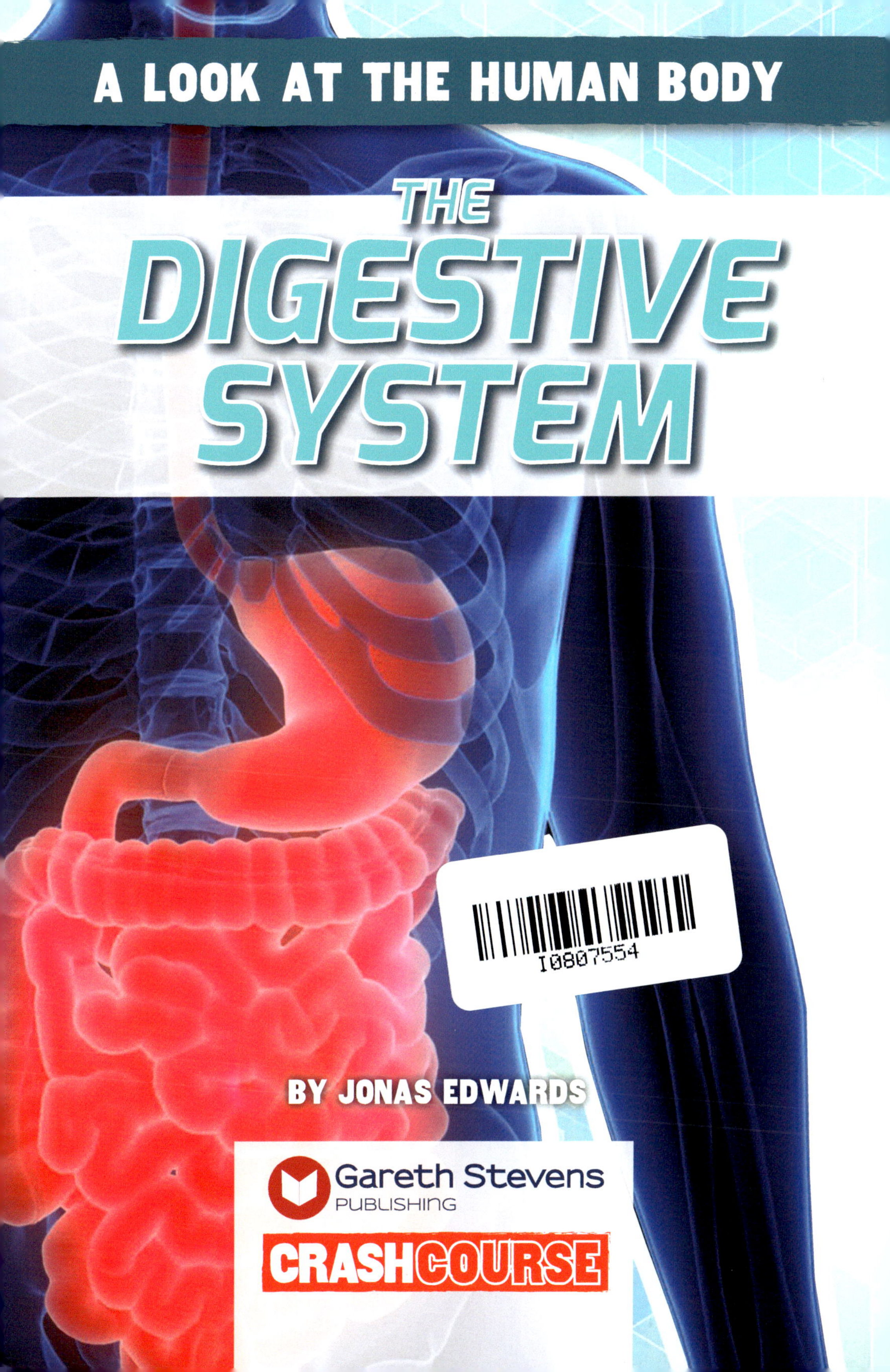

BY JONAS EDWARDS

Gareth Stevens
PUBLISHING

CRASHCOURSE

Please visit our website, www.garethstevens.com. For a free color catalog of all our high-quality books, call toll free 1-800-542-2595 or fax 1-877-542-2596.

Library of Congress Cataloging-in-Publication Data

Names: Edwards, Jonas, author.
Title: The digestive system / Jonas Edwards.
Description: New York : Gareth Stevens Publishing, [2022] | Series: A look at the human body | Includes index.
Identifiers: LCCN 2020014263 | ISBN 9781538266113 (library binding) | ISBN 9781538266090 (paperback) | ISBN 9781538266106 (6 pack) | ISBN 9781538266120 (ebook)
Subjects: LCSH: Digestive organs—Juvenile literature.
Classification: LCC QM301 .E39 2022 | DDC 612.3—dc23
LC record available at https://lccn.loc.gov/2020014263

First Edition

Published in 2022 by
Gareth Stevens Publishing
29 E. 21st Street
New York, NY 10010

Editor: Kate Mikoley
Designer: Andrea Davison-Bartolotta

Photo credits: Cover Magic mine/Shutterstock.com; background textures Yuliia Petrenko/Shutterstock.com; p. 5 (top) Inside Creative House/iStock/Getty Images Plus/Getty Images; p. 5 (bottom) Olivar Eltinger/Corbis/Getty Images Plus/Getty Images; p. 7 (top) Chris Stein/DigitalVision/Getty Images; p. 7 (bottom) SDI Productions/iStock/Getty Images Plus/Getty Images; p. 9 Jose Luis Pelaez Inc/DigitalVision/Getty Images; p. 11 (top) Flashpop/Stone/Getty Images; p. 11 (bottom) SCIEPRO/Science Photo Library/Getty Images; p. 13 (main) ttsz/iStock/Getty Images Plus/Getty Images; p. 13 (background) Olga OleNN/Shutterstock.com; p. 15 Nerthuz/iStock/Getty Images Plus/Getty Images; pp. 17, 23 Sebastian Kaulitzki/Science Photo Library/Getty Images; p. 19 Pixologicstudio/Science Photo Library/Getty Images; p. 21 ChiccoDodiFC/iStock/Getty Images Plus/Getty Images; p. 25 coward_lion/iStock/Getty Images Plus/Getty Images; p. 27 (inset) Caroline Arquevaux/Corbis Documentary/Getty Images Plus/Getty Images; p. 27 (main) RobertsGalleries/iStock/Getty Images Plus/Getty Images; p. 29 Science Photo Library/Getty Images; p. 30 Vecton/Shutterstock.com.

Printed in the United States of America

CPSIA compliance information: Batch #CSGS22: For further information contact Gareth Stevens, New York, New York, at 1-800-542-2595.

CONTENTS

Words in the glossary appear in **bold** type the first time they are used in the text.

BREAKING THINGS DOWN

Your body gets the **nutrients** it needs from foods you eat. But how do the turkey sandwich and apple you ate for lunch turn into something your body can **absorb** and take in nutrients from? That's where the digestive system comes in!

MAKE THE GRADE

Digestion is the breakdown of food inside the body. It turns food into matter your body can use!

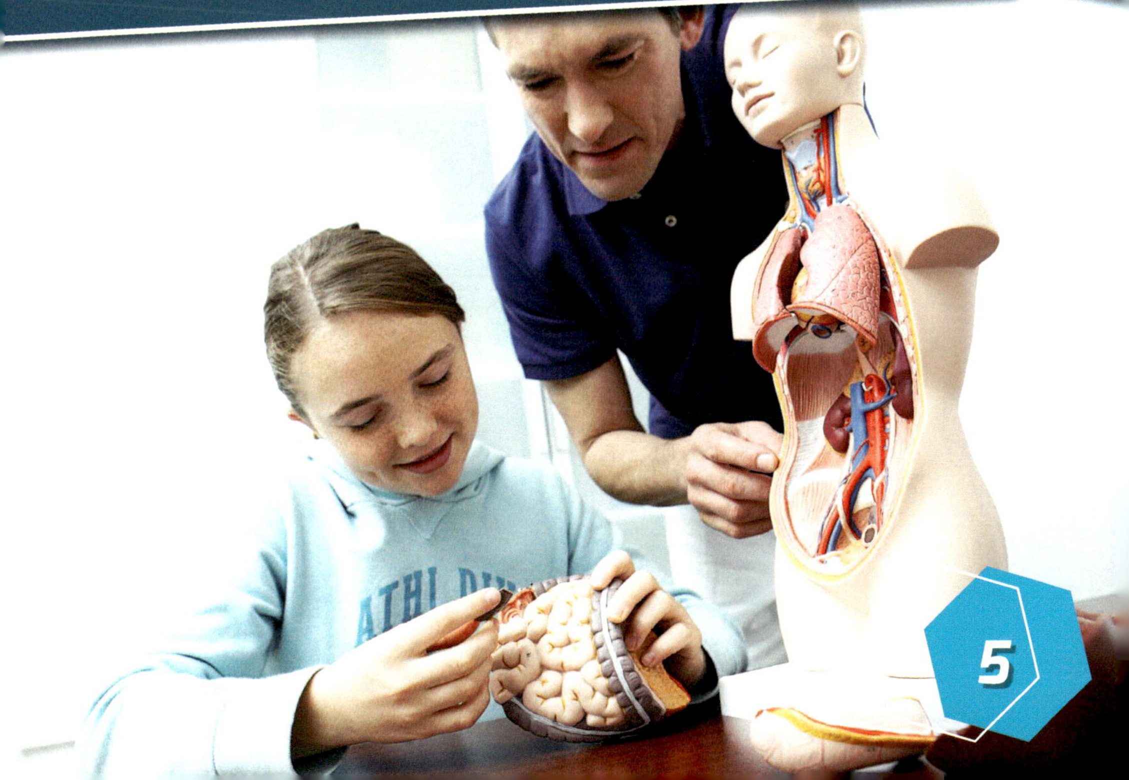

After you eat or drink something, it travels through a series of **organs** in your body that make up the digestive tract. Along the way, the food and drink are broken down in a way that allows nutrients to be absorbed into your blood.

MAKE THE GRADE

If you laid all the parts of an adult's digestive tract in a straight line, it would be about 30 feet (9 m) long. That's almost as long as a school bus!

MOVEMENT IN THE MOUTH

The digestive tract starts with your mouth, but the digestive system starts working even before you eat anything. Saliva, or spit, can form in your mouth when you smell, see, or even think about food! Once you eat something, saliva starts to break it down.

MAKE THE GRADE

Your teeth break food down into smaller pieces, but it's saliva that makes the food in your mouth mushy.

Your tongue pushes food around in your mouth, helping to soften it. It then helps you swallow the chewed-up food and saliva. Together, the food and saliva are called bolus. The food moves into the throat and then enters the esophagus.

MAKE THE GRADE

The esophagus is a **tube** that brings food from your throat to your stomach. It's about 10 inches (25 cm) long.

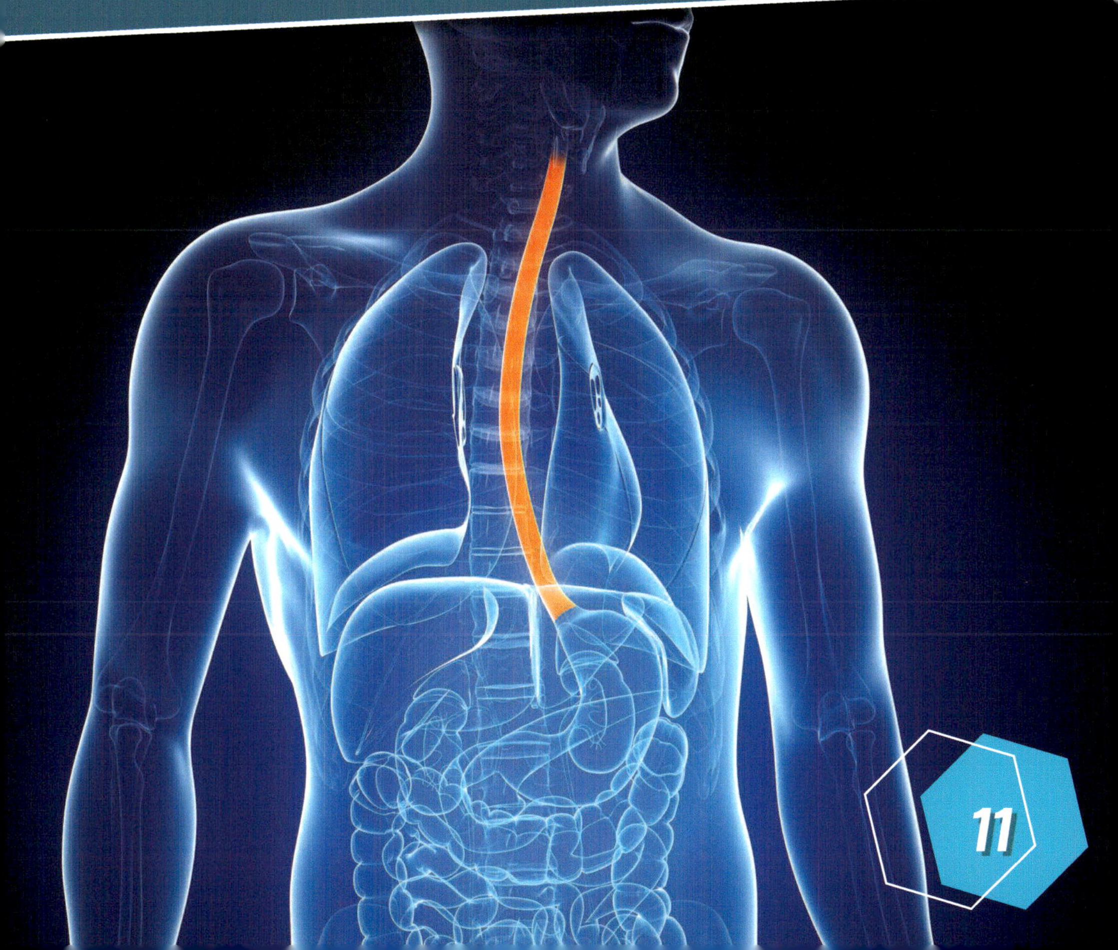

THE EXCELLENT ESOPHAGUS

The esophagus is commonly less than 1 inch (2.5 cm) wide. Rather than food just sliding down, **muscles** in the walls of the esophagus squeeze the food gently through. This commonly takes less than 10 seconds.

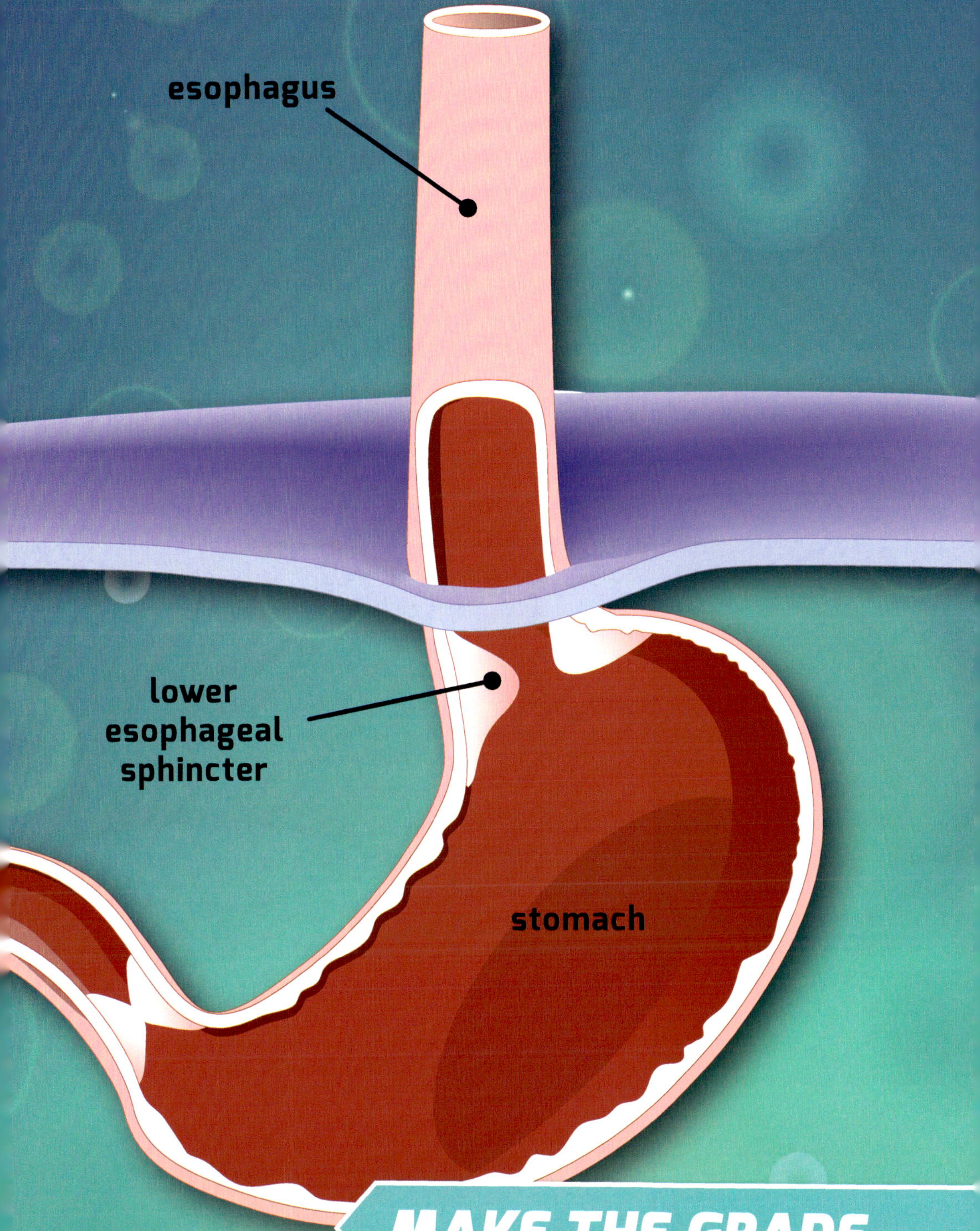

MAKE THE GRADE

At the end of your esophagus is a muscle called the lower esophageal sphincter. When it relaxes, food moves out of the esophagus and into the stomach.

YOUR SUPER STOMACH

The stomach is a sac shaped sort of like the letter J. All the foods that come through the esophagus get mixed and mashed into smaller pieces here. They also get mixed with **fluids** in the stomach called gastric juices.

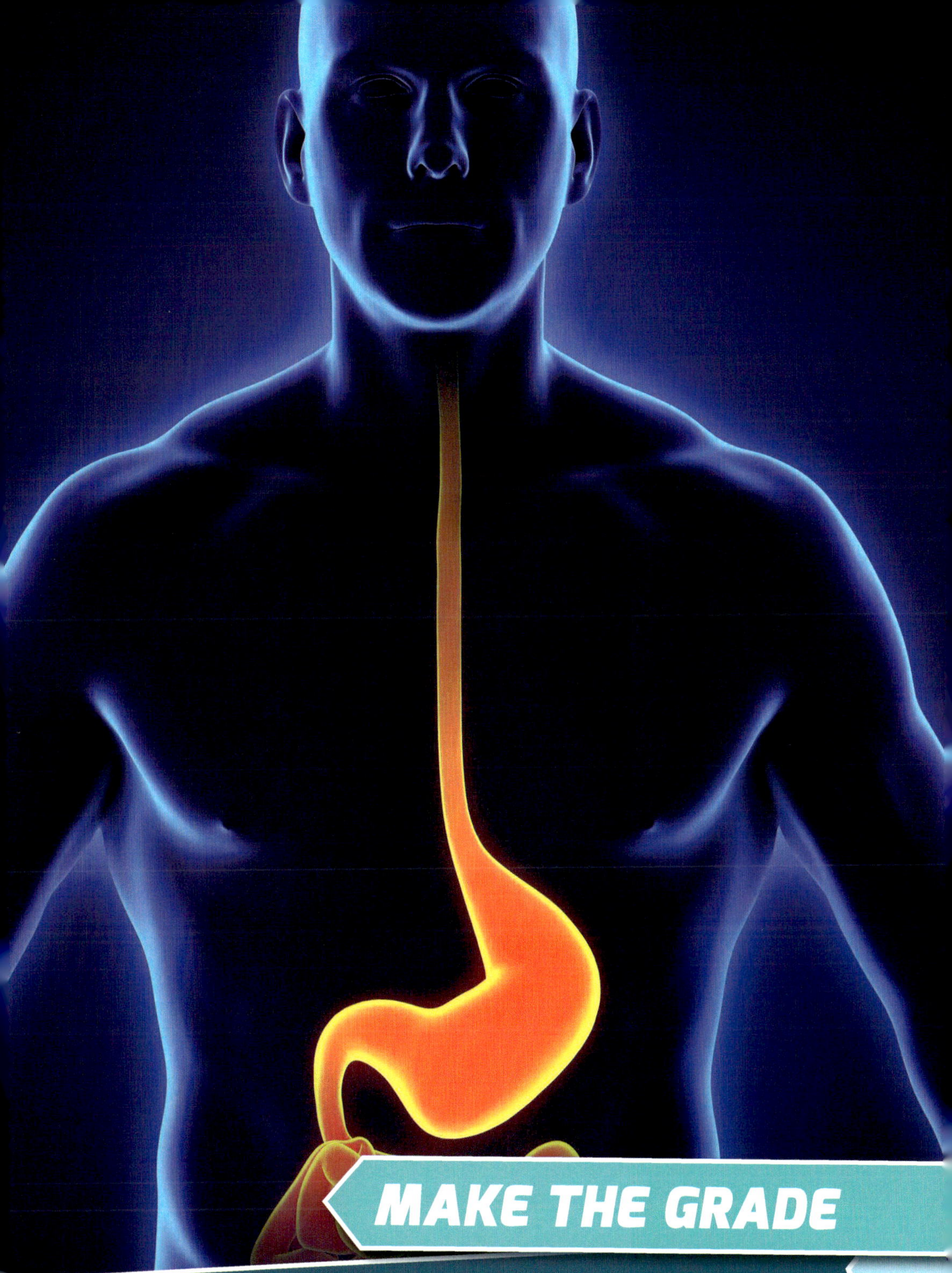

MAKE THE GRADE

Gastric juice is a watery acid given off by glands in your stomach. It helps break down food. It also kills any bacteria that may have been in your food.

SMALL BUT MIGHTY

From your stomach, the mix of partly digested food, called chyme, moves to your small intestine. This is a long tube that folds and bends many times, fitting into the space below the stomach. It breaks up the food even more.

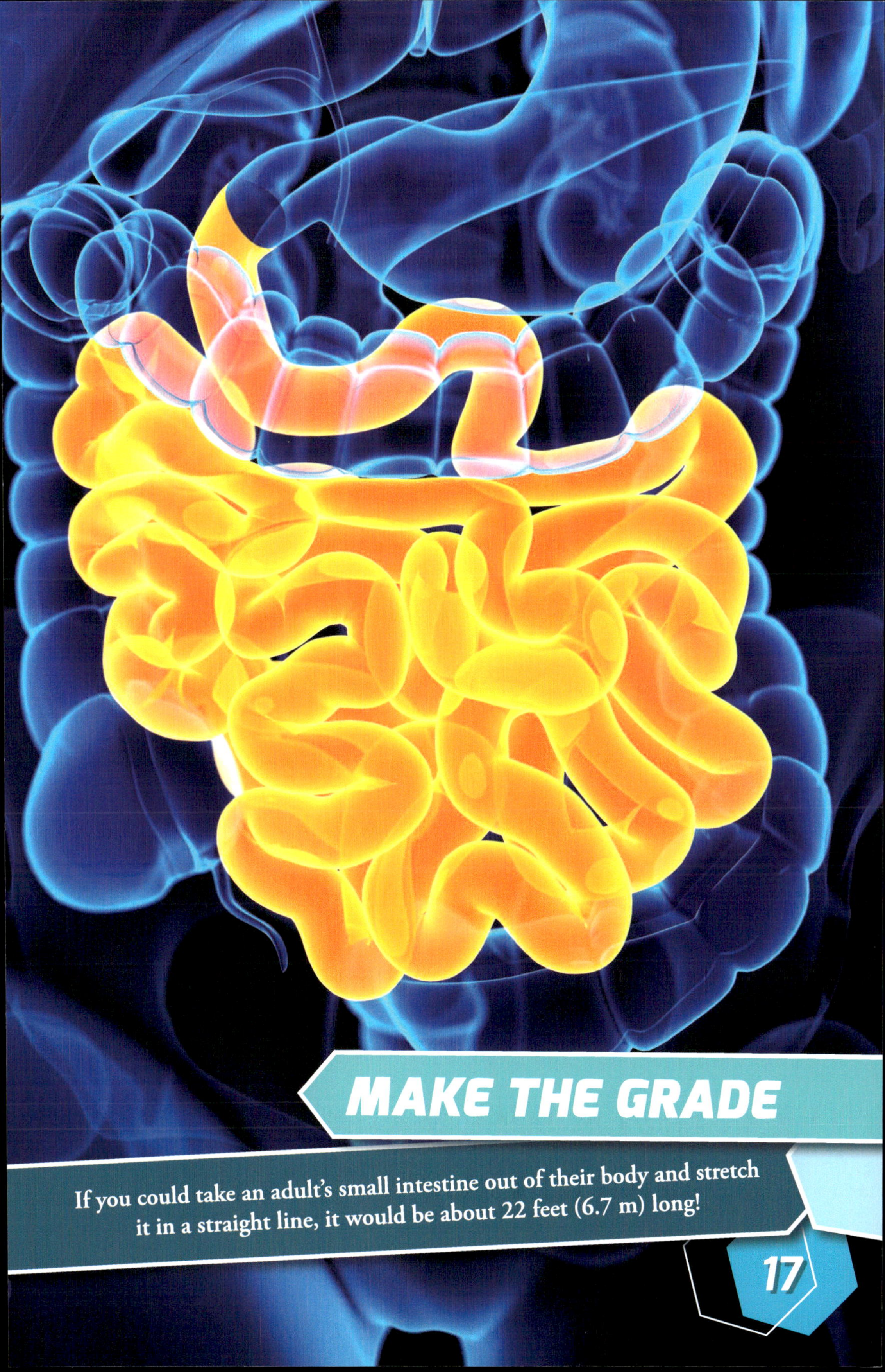

MAKE THE GRADE

If you could take an adult's small intestine out of their body and stretch it in a straight line, it would be about 22 feet (6.7 m) long!

In the small intestine, your body absorbs nutrients. An organ called the pancreas sends a special fluid there to help digest fat and **protein**. Another organ, called the liver, produces a fluid called bile. It helps the bloodstream absorb fats and **vitamins**.

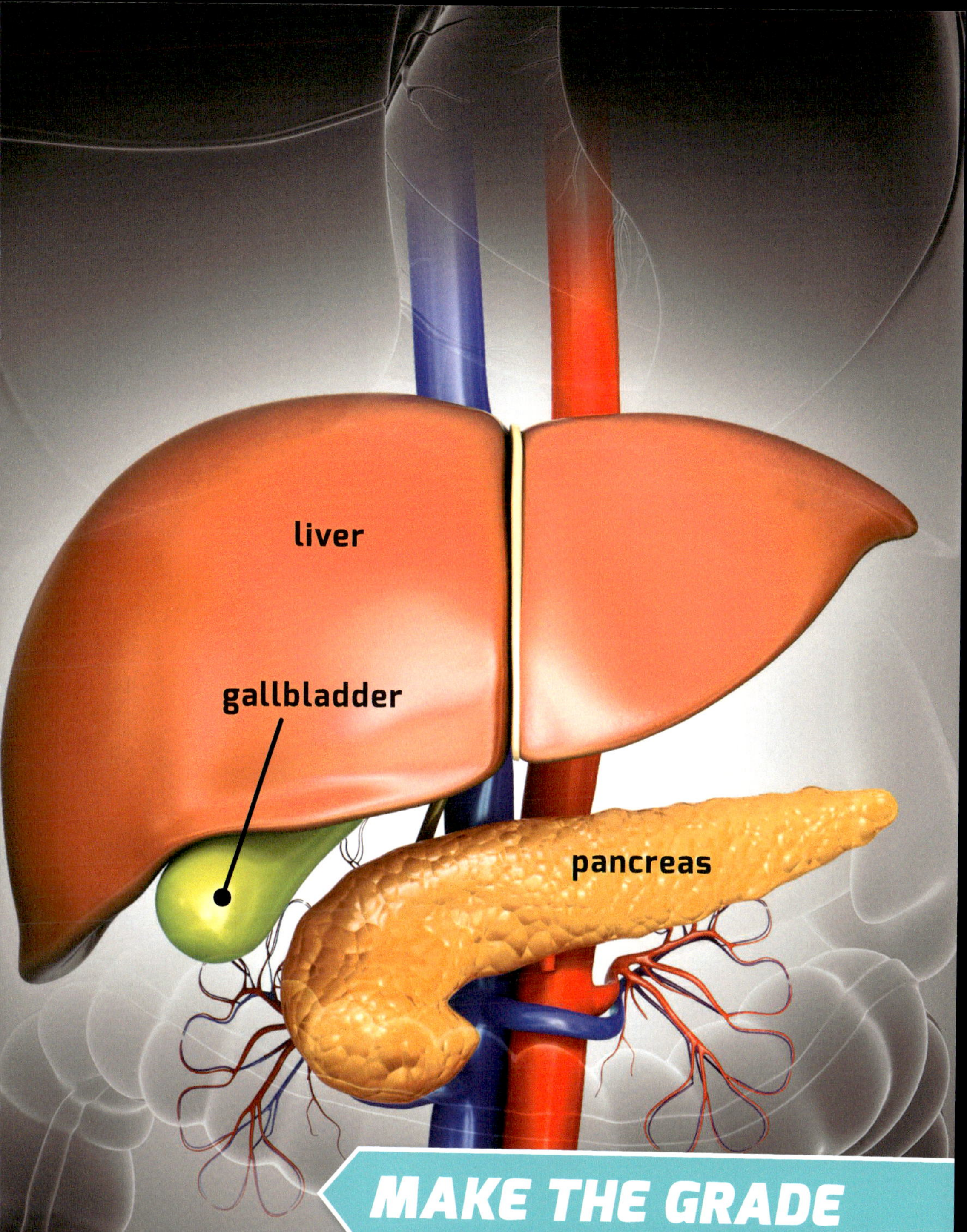

MAKE THE GRADE

Bile is stored in an organ called the gallbladder until it is needed.

LIFE OF THE LIVER

From the small intestine, blood that has absorbed nutrients goes to the liver. The liver removes any harmful matter or waste that may still be in this blood. Then, the liver sends the nutrients to other parts of the body as needed.

MAKE THE GRADE

The harmful matter removed by your liver leaves your body either through poop or pee!

LARGE AND IN CHARGE

After your body takes in nutrients from the food you ate, what's left is waste that your body needs to get rid of. This leftover matter goes into the large intestine. This tube is wider but shorter than the small intestine.

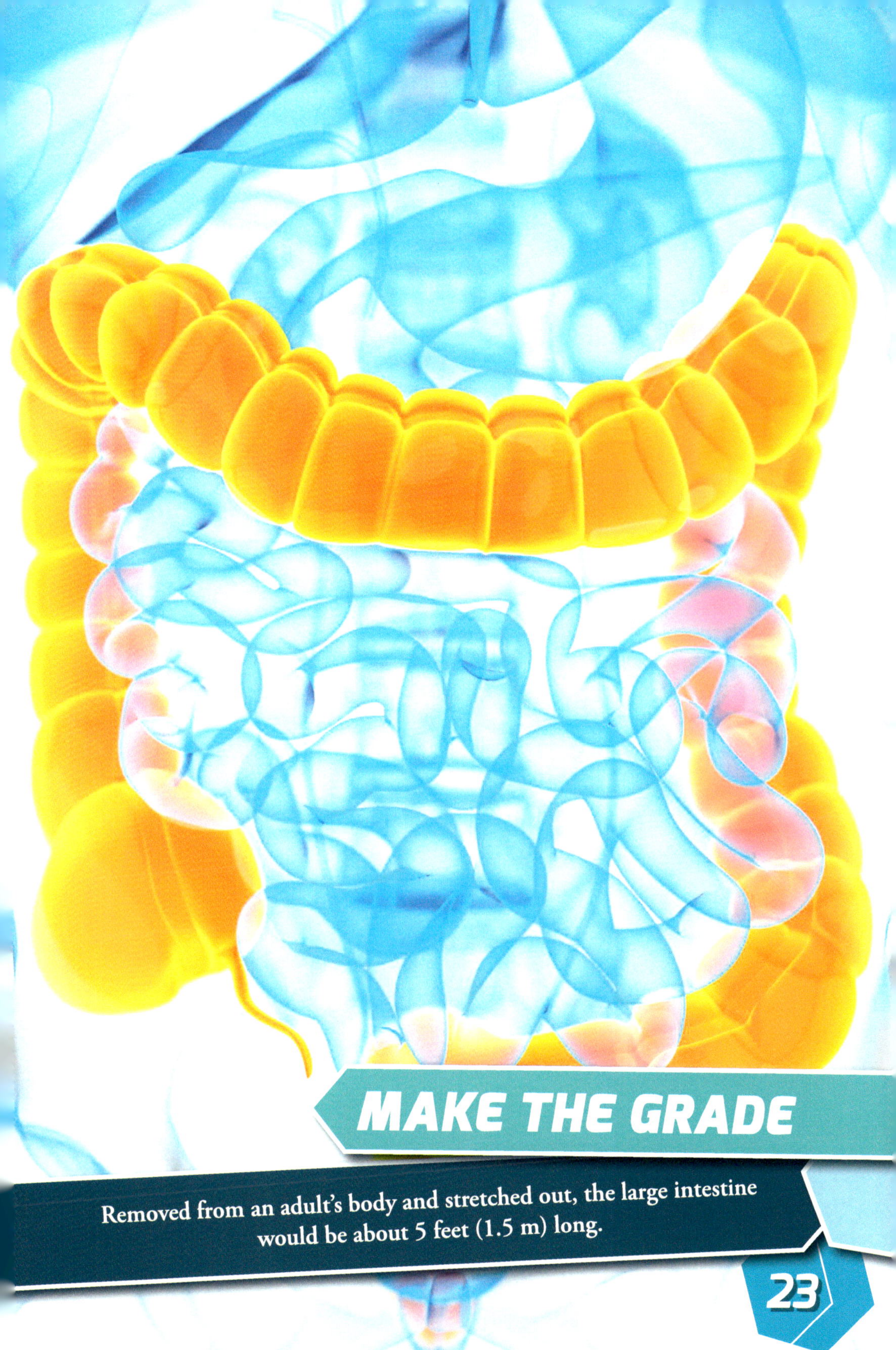

MAKE THE GRADE

Removed from an adult's body and stretched out, the large intestine would be about 5 feet (1.5 m) long.

Waste goes through a part of the large intestine called the colon. Here, the body has one more chance to absorb any water or nutrients still left in what started out as your food. Without much water, the waste becomes a solid matter—poop!

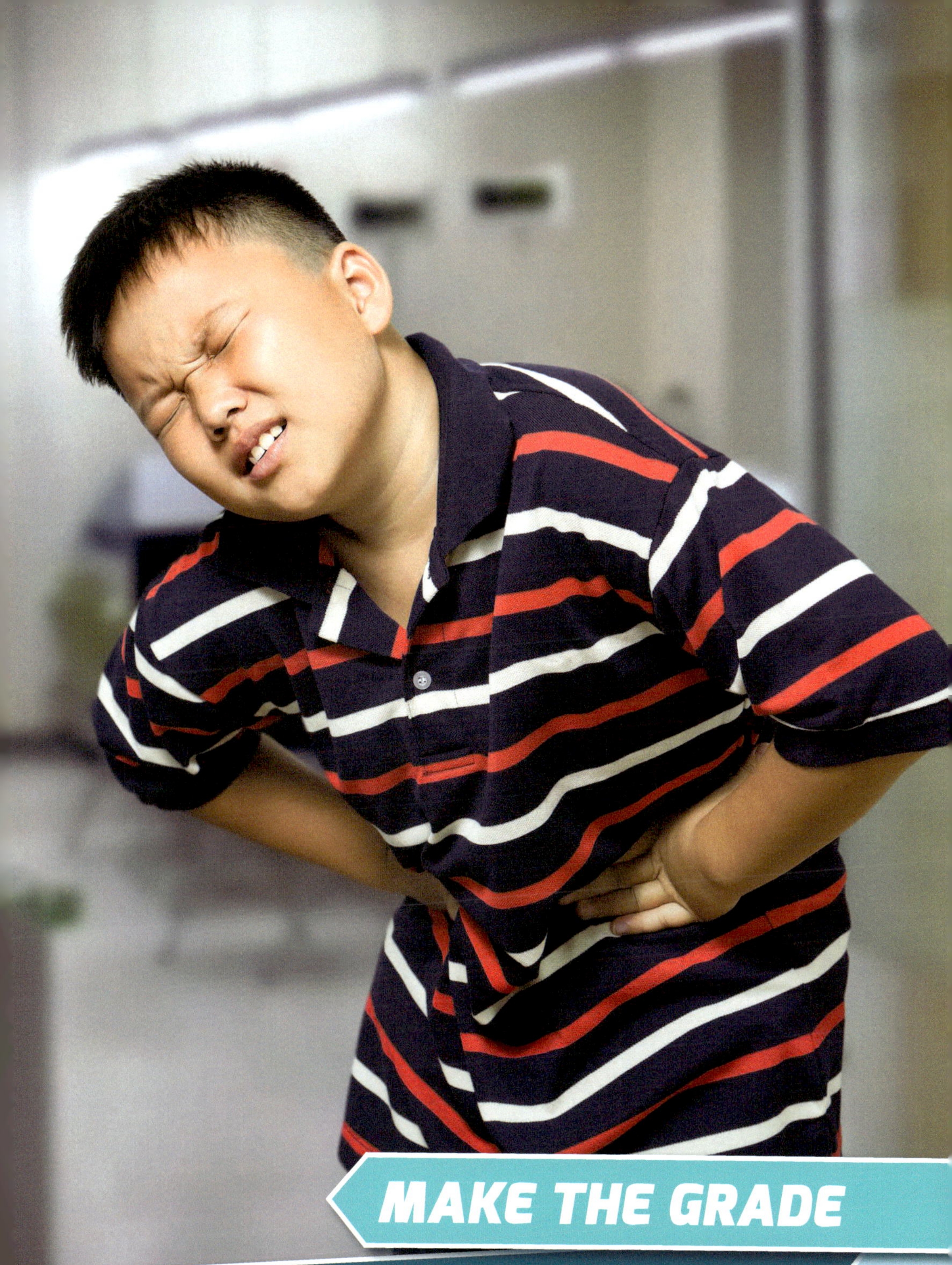

MAKE THE GRADE

Sometimes your poop might be watery instead of solid. This is fairly common and can happen if you're sick or ate something bad. If it happens a lot, it could be a sign of a colon problem.

WHERE'S THE WASTE?

From the large intestine, waste goes into the rectum. This is where the poop stays until you have to go to the bathroom. When you do go, the waste leaves your body through a part called the anus.

MAKE THE GRADE

After you eat something, it could take anywhere from one to five days until your body has fully digested it and gotten rid of the waste.

FOOD ON THE MOVE

You likely don't think too much about what happens to your lunch after you eat it. Maybe now you will! Your first bite is just the start of your food's journey. Your digestive system is always working to help you stay well.

MAKE THE GRADE

Like the rest of your body, it's important to keep your digestive system healthy. One way you can do this is by drinking plenty of water.

PARTS OF THE DIGESTIVE SYSTEM

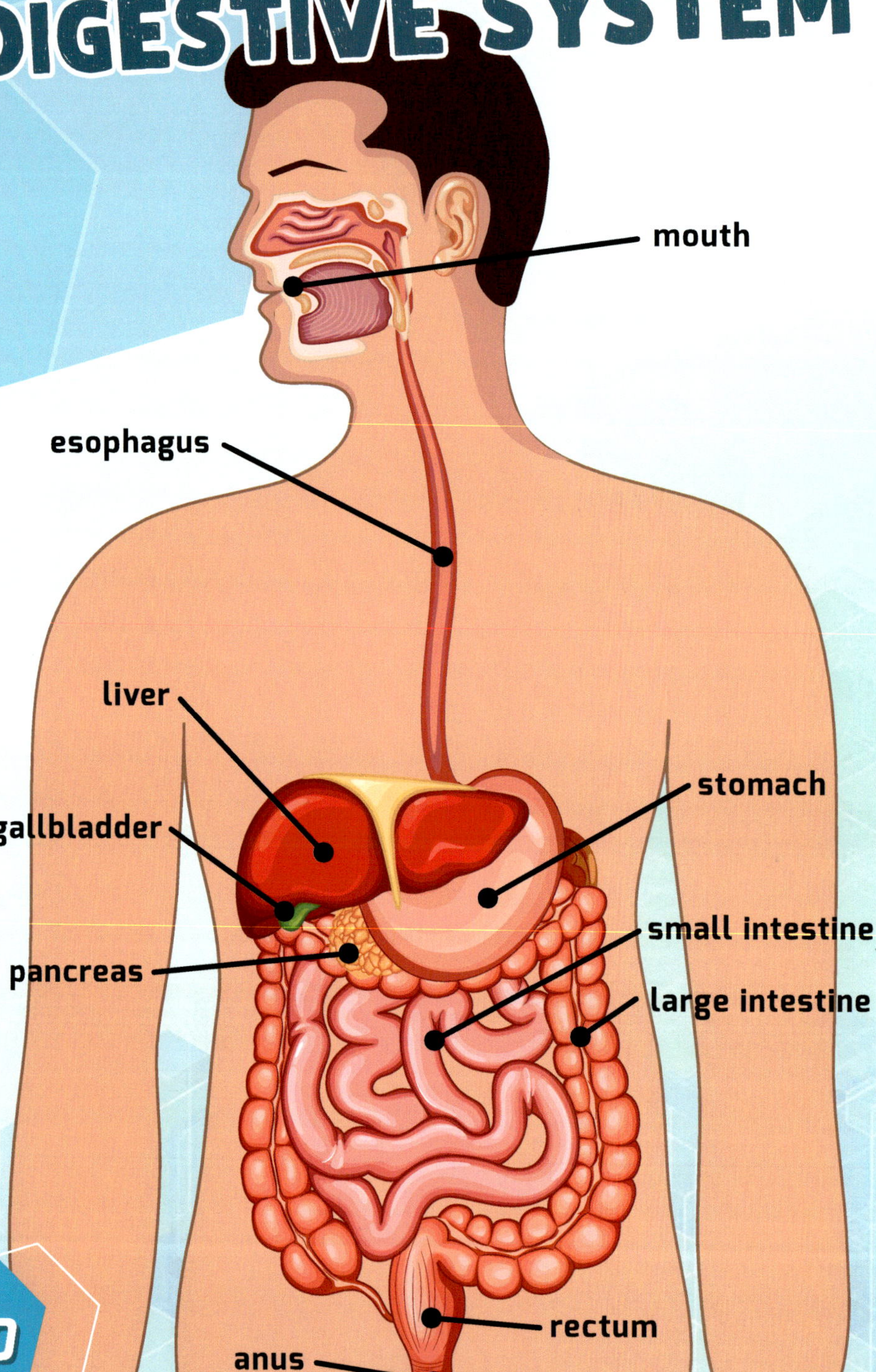

GLOSSARY

absorb: to take in

acid: a liquid that breaks down matter

fluid: something that is watery and flows like a liquid

gland: a body part that produces something needed for a bodily function

muscle: one of the parts of the body that allow movement

nutrient: something a living thing needs to grow and stay alive

organ: a part inside an animal's body

protein: a nutrient in many types of food that the body uses to grow, repair tissues, and stay healthy

tube: a thin, long, hollow part of an animal that something can move through

vitamin: a natural matter that is often found in foods and helps a body be healthy

FOR MORE INFORMATION

BOOKS

Farndon, John. *Stickmen's Guide to Your Gurgling Guts.* Minneapolis, MN: Hungry Tomato, 2018.

Raij, Emily. *My Digestive System: A 4D Book.* North Mankato, MN: Pebble, 2019.

WEBSITES

Digestion
www.dkfindout.com/us/human-body/digestion/
Read more about digestion on this interactive website.

Your Digestive System
kidshealth.org/en/kids/digestive-system.html
Head to this page to learn even move about your digestive system.

INDEX